What we do at LaserPronet

- Screening Exams
- Professional Development Courses
- Professional Growth Plans
- Certifications

Copyright © 2018

Sukuta Technologies, LLC

All rights reserved

Laboratory Manual for Laser Fabrication and Factory-Level Tuning and Performance Tests

Table of Contents

	Topic	Page
1.	Laser Technician III	4
2.	Introductory laser lab notes	5
3.	Components handling, inspections and cleaning	7
4.	Understanding laser mirrors' spectrographs	13
5.	Understanding solid-state laser rods' spectrographs	16
6.	Laser fabrication/alignment	17
7.	Lab 1. Laser Resonant Cavity Fabrication	36
8.	Lab 2. Laser Pump Output Wavelength Tuning	41
9.	Threshold, Slope efficiency, and Warm-up Time	46
10.	Lab 3. Collimation of Diverging Laser Beams	50
11.	Lab 4. Laser Threshold and Slope Efficiency	59
12.	Lab 5. Laser Beam Harmonics: Temp-tuning and Conversion efficiency	62
13.	Power supplies and wall-plug efficiency	65
14.	Lab 6. Laser Lower Supply and Wall-plug Efficiency	67
15.	Laser Relative Stability Fluctuations and the ISO Standard of Stability Tests (ISO 11554)	71
16.	Useful Lifetime, Burn-in and Footprint	74
17.	Self-Test	75

1. Laser Technician III

Laser Technician III Job Description: Level III Technicians assemble, align, burn-in, test, and tune/troubleshoot laser heads until they meet all performance specifications.

Environment: Manufacturing

Requirements: Must understand the fundamentals of solid state laser technology, accessories and support systems. Experience aligning laser systems and cavities/resonators and using photonics test equipment is a must. Also, ability to follow prescribed written procedures, directions and strict adherence to best laser lab and manufacturing practices are required.

Minimum Educational Requirements: Candidate must hold a certificate/degree in Laser/Electro-Optics Technology or related discipline

2. *Introductory Laser Lab Notes*

The following applies to all experiments/labs in this course and program

a. Replicate and fill out the data tables given in your lab notebooks. You can also modify them to match your style better.
b. Record all your observations in your lab notebook during lab-time.
   - Data transcribing after the fact/experiment violates intellectual property protection guidelines.
   - Comment on your findings in your lab notebook.
   - After the lab session sign your lab notebook at appropriate places and have the instructor sign it too before you leave the lab.
c. Clean-up your work area and re-shelve the equipment when done with your lab work/ experiment.

a. Warnings When Operating Lasers

d. Chiller must be running before laser is fired, where applicable.
e. Appropriate laser eyewear must be worn before the laser is fired.

f. Power must be attenuated to zero then slowly increase it until a signal is detected/observed.
g. Make sure to know the operation and limitations of the laser analyzer you are using to avoid damaging it and getting the wrong measurements.

h. When applicable, the chiller must run for at least 15 minutes after the laser is turned off to ensure that the laser rod does not get damaged.
i. You must put on appropriate eyewear before a laser is turned on.

b. Results Reporting

Always prepare a

1. Test Procedure,
2. Oral and
3. Poster Presentation

after every lab exercise.

3. Laser Components Handling, Inspections and Cleaning

3.1. Optics Handling

Extreme care and caution should be taken when handling optics in order not to scratch/damage them. Always wear gloves/finger cots and place all optics on notes padded surfaces. All optics items must be in protectives cases or bags e.g. bagged air-bubbled plastic bags.

Compromised/damaged coatings on optical entities will not perform as desired or specified. It is therefore imperative that we can measure the performance of laser optics and mirrors as way to ensure proper performance before installing them in a laser.

3.2. Why Inspect and Clean Optics

Damages and dirt on optics will degrade laser/laser system's performance. Optimum performance and maximum lifetime are achieved when optics are clean and damage-free.
Both dirt and damages may alter the specified performance of a given optics.

Contamination/dirt scatters light and absorbs laser energy leading to hot spots. Any hot spots may eventually lead to coating degradation and subsequent failure. In any case, there is always a risk to damage optics if is cleaned too often and unnecessarily. It is therefore important that all optical assembly work is done in a controlled clean environment and be handled properly. Both components and assembled systems must be stored in clean rooms. Finished components must be sealed sufficiently enough to disallow the inflow of dirt always.

3.3. Inspection and Cleaning Tips

Table 3.1 Minimal Equipment Needed

#	Item	Purpose
1	Bright Light	Illumination
2	Clean room tissue pad	placing optics
3	Compressed air/nitrogen	Blow cleaning
4	Gloves/finger cots	Protect optics from bare hands
5	Hemostat	for holding lens tissue
6	Package of lint-free lens tissue	For swipe-cleaning optics
7	Solvent (methanol/acetone/ isopropyl alcohol) with eyedropper	To be applied on lint free tissue
8	Swabs/Q-tips	For swipe-cleaning optics, particularly in hard to reach places
9	Tweezers	for holding small optics
10	a. Visual Bare Eye b. Loupes/Magnifying Glasses c. Microscopes d. Camera e. Cameras and Microscopes f. Interferometers g. Spectrometers	Optics dirt/damage viewing
11	Benchtop	

- Make sure that you, the technician, and the area you work in are not sources of contamination/damage
- Start with a clean bench top, and if possible with edge restraints around the area to prevent the optics from rolling away and onto the floor
- Place clean room tissue pad on bench-top
- Place your optics on this soft-landing pad
    - Also dropped optics will not take maximum impact if dropped.
- Wear gloves. As you wear gloves, hold them from the wrist edge to avoid contamination
- Gloves also protect optics from being contaminated by your hands
- Gloves also protect your hands from solvents.
- Minimize the cleaning of optics to minimize the probability of damaging them
- If an optics is dirty, try to use none contact cleaning methods first such as gas blower, such as compressed nitrogen and air.
- If none contact methods do not workout escalate to using contact methods
- Use the proper equipment and accessories when cleaning optics
- Always use fresh dry solvents everyday as old ones may have picked up atmospheric moisture.
- When cleaning optics make sure to observe the "One-wipe Rule" to avoid the spread of contaminants around and possible damage of the optics being cleaned
- Make sure all optics surfaces are clean otherwise the dirt will burn the optical coatings.

- Avoid any contact with the inside of the laser cavity.
- Constantly blow compressed air/nitrogen, if available, in the cavity to keep contaminates out.

3.4. Potential Human-induced Optics Damage
- Holding optics with bare hands will lead to optics damage/failure
    - You must always wear finger cots or gloves when handling and cleaning optics
- Carelessly handling tools like tweezers, hemostats etc. while working with optics could lead to optics damage/failure.
- Humans can lower optics damage threshold and performance due to the trafficking of
    - dirt
    - lint
    - sweat/body fluids
    - Fingerprints
    - body oils
    - hair
  onto optics
- Touching any part of an optics accessory that contacts optics will lead to optics damage/failure
    - For example, holding the "fingers parts" of gloves with bare hands could lead to optics damage/failure.
- Failure to detect and reject optics with voids such as digs and scratches, and lint, dirt, sweat, fingerprints etc. will eventually lead to optics damage.
- Failure to remove from the workplace/area airborne particles residue from out-gassing of components will eventually lead to optics damage
- You should never place optics on a bare optical bench top but on a clean-room tissue pad
- A clean bench-top will prevent an optic from being contaminated

- It is a good laser lab practice grab optics along the edges, not the its face/aperture.
- Failure close/cap solvent containers would result in their degradation thus leading to poor cleaning performance.
- If optics is excessively and unnecessarily cleaned it has a high probability of being compromised.
- Failure to inspect optics before cleaning it could turn "good" optics into "bad" optics.
- Always using contact methods on optics that (a) did not need cleaning, or (b) could be cleaned using non-contact methods could lead to rapid degradation of the optics
- To minimize, and hopefully avoid, optics damage always starts non-contact optics cleaning methods first and then resort to contact methods last
- Cleaning optics using dirty tools could lead to the inclusion of optic cause damage and inclusion of foreign objects.

3.5. Optics Surface Quality Specifications

- US military specifications Mil-O-13830A "Optical Components for Fire Control Instruments ", is now the de-facto Optics Quality Inspection standard.

    - Scratch-Dig Notation is used to characterize defects in optics.
    - For example, 20-4 means that the maximum acceptable width of a scratch is 20 microns and the diameter is 40 microns for a dig
    - For example, a laser buyer may specify OC mirrors: 10-5 for surface S1 and 20-10 for surface S2
        - S1 is surface facing the active medium inside the laser cavity
        - S2 is surface facing away from the active medium
    - HR (High Reflector) mirrors: 10-5 for surface S1 and, 80-50 for surface S2
    - Beam-splitters: 30-20, Filters: 30-10 etc...

- Also Scratch-Dig Notation has equivalent letter designations (Mil-F-48616) as noted below.

3.6 Workforce Skills Desired
Ability to
1. Inspect optics and decide if it must be cleaned.
2. Decide which optics cleaning method is appropriate if the optics is dirty
3. **Clean the optics without damaging**

4. Laser Mirrors' Spectrographs

It is imperative that optics with the correct coatings are used in the resonator. Optical coatings allow laser mirrors to reflect [R] and transmit [T] specific wavelengths. For example, a HeNe laser, $\lambda=632.8$ nm, with an Output Coupler (OC) with a reflectivity of 95% will transmit (T) 5%. On the other hand, High Reflectors, HR, are assigned reflectivity values of 100% on any lasers.

In general

T+R+A=100%

Where A is the absorbance and we assume its zero (A=0) or negligible for all optics under analysis, so equation 1 is reduced to

T+R=100%
If we can therefore measure a laser mirror's transmission, T, at a specific wavelength we can go on to calculate its reflectivity, R, as follows

100-T=R

Note that generally reflectivity, R, is specified on optics while some spectrometers may only measure transmission, T, therefore it is left to the user to convert transmissions to reflectivity's as shown in equation 3.

Workforce Skills Required
Ability to inspect optics and decide if it
    a. must be cleaned and if
        i. Decide which optics cleaning method is appropriate if the optics is dirty
        ii. Clean the optics without damaging
        iii. has the proper optical coating transmissions at the wavelength(s) of interest.

Always follow company posted procedures and guidelines
Laser Mirrors' Spectrographs

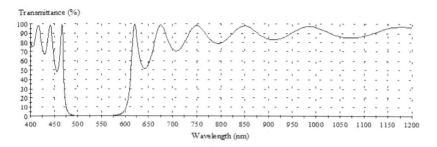

Figure 4.1. Mirror/Optics # 1

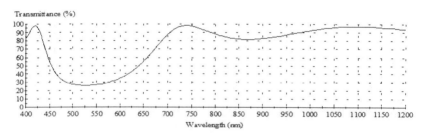

Figure 4.2. Mirror/Optics # 2

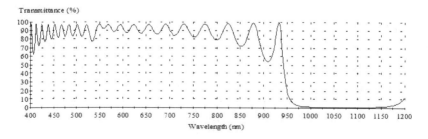

Figure 4.3. Mirror/ Optic # 3

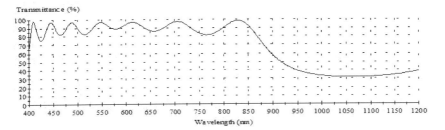

Figure 4.4. Mirror/Optics 4

Table 4.1. Laser mirror data

Optics #	Type of coatings	Coating materials	Coating Process	Design wavelength (nm)	Target Reflectance	Coating Design	No. Layers	Individual and total Physical Layer	Substrate material	Substrate Diameter	Refractive Indices (at design λ)
1	532 nm dielectric HR	Ta2O5 SiO2	IBS	532	99.8%	G.(LH)^11.A	22	L (SiO2): 89.97 nm H (Ta2O5): 62.18	fused silica	25.4 mm dia x 5mm	L (SiO2): 1.478 H (Ta2O5): 2.139 Sub: 1.4608
2	532 nm dielectric OC	Ta2O5 SiO2	IBS	532	74.2%	G.(LH)^4.A	8	L (SiO2): 89.97 nm H (Ta2O5): 62.18	fused silica	25.4 mm dia x 5mm	same as above
3	1064 nm dielectric HR	Ta2O5 SiO2	IBS	1064	99.7%	G.(LH)^11.A	22	L (SiO2): 181.27 nm H (Ta2O5): 128.96	fused silica	25.4 mm dia x 5mm	L (SiO2): 1.467 H (Ta2O5): 2.063 Sub: 1.4497
4	1064 nm dielectric OC	Ta2O5 SiO2	IBS	1064	69.9%	G.(LH)^4.A	8	L (SiO2): 181.27 nm H (Ta2O5): 128.96	fused silica	25.4 mm dia x 5mm	same as above

Source: Laser mirror spectrographs and associated data, courtesy of Bruce Perilloux of Coherent Inc. (March 2015)

5. Understating Solid-State Laser Rods' Spectrographs

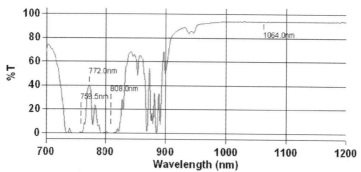

Figure 5.1. ND:YAG rod transmission spectrum in visible and NIR

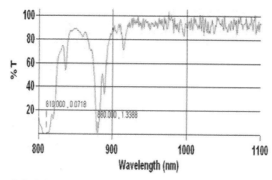

Figure 5.2. ND:YVO$_4$ rod transmission spectrum in NIR

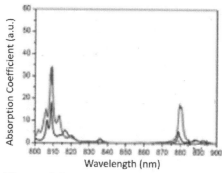

Figure 5.3. ND:YVO$_4$ rod absorption spectrum in NIR

6. Laser Fabrication/Alignment

6.1. Technician Workplan.
6.2. Laser Resonator Fabrication/Alignment
6.3. Laser Resonant Cavity Fabrication Example.

6.1. Technician Workplan.

Laser technicians in laser manufacturing companies are expected to build and test lasers. To be successful, they should be familiar with how a laser works in general, and how each specific component works as well, so that they have a better grasp of how they interact with each other to generate laser "light"/radiation. In addition, they should have hands-on skills to evaluate and assemble components and subassemblies and make them work together to make the laser "lase" for the first. Once the laser is working they should also be able to test and improve/optimize its performance. Highly skilled technicians (Tech IV) are also trusted with the task of troubleshooting lasers/laser systems.

- Electro-mechanical Laser Assembly
    - Laser Head
        - Mirrors
        - Intracavity components
        - Laser Pump/Laser Power Supply
    - Laser System
        - Alignment of laser beam through optical components of the system. The laser head is treated as a blackbox but a part of the system.

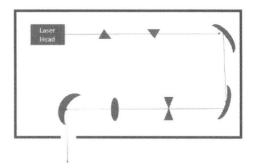

Figure 6.1. Depiction of a laser system

- The assembly may include soldering/mounting of electrical, optical and mechanical components.

  • Check for any broken electrical cables/wires

  • make sure all electricalwires/cable are connected properly.

  • Good solder joints are usually shinny while bad/cold ones are dull.

  • Tie wrist strap everytime you handle electronic circuits such as those found on Printed Circuit Boards (PCBs)

- Fabricate/Align then tune/optimize performance
- Beam Quality Tests and Measurements/Final Test
- Troubleshoot

6.2. Laser Fabrication/Alignment
    6.2.1. Laser Resonator Alignment Goals
    6.2.2. Laser Alignment Methods and Strategies
    6.2.3. Laser Resonator Alignment Set-up
    6.2.4. Some Rules of Thumb While Aligning a Laser
    6.2.4. To Lase or Not to Lase
    6.2.5. Laser Resonator Output Optimization

6.2.1. Resonator Alignment Goals

The general goal of aligning optical components in a laser resonator/system is to ensure that all the optical elements within it share a common optical axis. Specifically, a laser will not lase, or perform optimally unless the HR and OC (and all intracavity components) share a common optical axis as shown below.

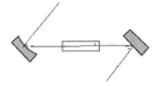

Figure 6.2. Misaligned Cavity

Figure 6.3. Aligned Cavity

Despite precise blueprints, fabricated components and subassemblies cannot perfectly match them because they have tolerances, so the final products will deviate from them. So, alignments also correct for deviations from idealized blueprints.

a. Why Laser Resonator Alignment?
   i. To make a laser resonator lase for the first time
   ii. To optimize/change laser power levels
   iii. To improve/change beam parameters e.g. M^2, size, ellipticity/symmetry, divergence etc., such as on the laser test list

b. Why Laser/Optical Systems Alignments?
   i. To condition laser beam
   ii. To improve/change beam parameters e.g. M^2, size, ellipticity/symmetry divergence etc., such as on the laser test list
   iii. To suppress laser beam aberrations in an optical train/system
   iv. To guide and position laser beams

6.2.2. Laser Alignment Methods and Strategies

- Laser resonator fabrication will generally involve a "walk-through" of an external beam to pre-trace the path beam of the laser being built.
- A low power laser alignment beam is generally used for safety reasons.
  - Typically, a HeNe laser is used and/or a laser whose output beam wavelength is identical to the one being manufactured/fabricated

Laser Resonant Cavity/Head Alignment Procedures.

- Active medium "Thermal lens-centric" positioning.
  - This ensures that the beam passes through the optical axis of the thermal lens created by a solid-state active medium if not uniformly pumped or cooled.

- Aligning/HeNe laser beam coarse alignments
  - Dispersion effect differences between aligning beam and "lased" beam create dispersion effects discrepancies. Such alignment is therefore coarse and can/should be improved.
- Fundamental beam fine alignments
  - If coarse laser resonant cavity alignment are performed first using a laser beam different from that generated by the cavity itself, it would be necessary to optimize the laser performance by re-aligning the cavity using its self-generated laser beam.

6.2.3 Laser Resonator Alignment Set-up

6.2.3.a. Observing Best Practices
6.2.3 b. Fabrication in Cleanroom Environment
6.2.3.c. Typical Equipment Needed
6.2.3.d. Some Good Rules of Thumb Before Aligning a Laser

6.2.2. a. Observing Best Practices.
  - ✓ Always consult your Manufacturing or Product Engineer before working on a laser and/or if the task at hand is not clear enough or beyond your grasp.

  - ✓ Make it a habit to review your Good Laser Lab and Manufacturing Practices (GLLMP) guidelines before setting-up your alignment equipment. For example
    - Laser Safety
    - Electrical Safety

- Optics Handling, Inspection, and Cleaning
    ✓ Clean-up your work area with alcohol, or comparable, wipes before using and during work session if necessary
- ESD
    ✓ It's always best to be wrist strapped to avoid ESD
    ✓ Note that plastic containers accumulate charge so use ESD bags.
    ✓ Also, metal containers do not accumulate charge, and if they do they can discharge it particularly if the optical bench they are on is grounded.
    ✓ Metal tools, such as screwdrivers, can be ESD-proof if sat on/wrapped in foil

6.2.3. b. Fabricating Cleanroom Environment

Laser alignment/fabrication is generally performed in a clean environment called cleanrooms. Proper gowning is necessary to ensure that the laser and its components are not contaminated.

The lower the cleanroom class number the cleaner it is, for example a Class 100 Cleanroom is supposed to be cleaner than a Class 500 Cleanroom. Less sensitive lasers are fabricated in clean areas, with less stringent requirements.

6.2.3. c. Typical Equipment Needed
i. Optical work bench/optical breadboard
   a. Ideally with earthquake braces if in a a region susceptible to seismic activity.
ii. Electrically grounded for ESD protection.
iii. Appropriate screws and screwdrivers
iv. Appropriate tools for electronic tuning

- a. Always allow at least 30 seconds for tuning effect to kick in before moving to the next step.
- v. Alignment Laser – must be low power
- vi. Optics Inspection and Cleaning Kit with bright lighting
- vii. Safety eyewear for alignment laser and the laser being fabricated. [*If you are unable to see the laser beam and/or read your instruments talk to your Laser Safety Officer (LSO). You should never take off your protective eyewear while the laser is on.*]
- viii. Ruler
- ix. At least 2 beam steering mirrors
- x. Laser mount
- xi. At least 2 apertures/iris diaphragms
  Laser beam target
- xii. Lens tissue and/or invisible beam viewer if applicable.
- xiii. Power Meter
  - a. It is always a good practice to specify distance from the power meter beam sensor/detector to the laser for consistence in your output readings
  - b. Always "zero-out" ambient power reading off the power meter before taking any power measurements of a laser i.e. calibrate for ambient light.
- xiv. Optional/Nice to have equipment
  - a. Laser Profiler, if needed
  - b. Spirit level/bevel

Some Good Rules of Thumb Before Aligning a Laser

- ✓ Make sure breadboard is level.
  - o Check using instruments such as a spirit level.
- ✓ Inspect all components for cleanliness before assembling, and clean if needed.
  - o Make sure all optics and surfaces in the laser are clean otherwise the dirt will burn the optical coatings.
  - o When you clean optics do not leave solvent residue/streaks.

- ✓ Avoid any contact with the inside of the laser cavity with your bare hands now and throughout the alignment process.
- ✓ Hook-up the power supply but do not turn it on yet until after you are done with the alignment
- ✓ Install beam stopper in beam path
- ✓ Put on the prescribed laser alignment eyewear before turning on the alignment laser.
- ✓ Install the chiller/heat exchanger/any designated cooling system.
    - o If chiller/heat exchanger/or any liquid-based cooling device check leaks

6.2.3.d. Some Rules of Thumb While Aligning a Laser

6.2.3.d. 1. Typical Intracavity Components

6.2.3.d. 2. Good Alignment Practices

The standard laser industry practice to control and guide an alignment beam is achieved by using x-y alignment mirrors. This process is also referred to as "Walking the Beam".
- ✓ Keep your beams enclosed/contained within your optical bench/work area for the safety of people around you.
- ✓ Keep the inside of the resonator and the components to be assembled clean.
    - o Constantly blow compressed air/nitrogen in the cavity to keep contaminates out.
    - o Avoid contact with the internal parts by wearing finger cots/gloves.
- • When aligning a laser, an alignment laser beam is guided to enter through one end of the laser, e.g. through OC, to the other laser mirror, in this case HR, and must traverse the same path on its way back to the alignment laser

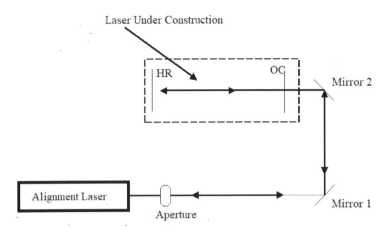

Figure 6.4. Typical laser resonator alignment set-up.

- Component by component alignment is performed aggregating into the final system alignment.

6.2.3.d. 1. Typical Intracavity Components

1. HR
2. OC
3. Pump/Lamp, another laser.
4. Active media/Nd:YAG, Nd:YVO4, ND:YLF etc..
5. Q-switches/ AOM or E-O
6. Harmonic crystals e.g. KTP, LBO. BBO etc..
7. Lenses/ focusing, collimating lenses etc...
8. Windows/ Brewster and pickoff windows etc..
9. Etalons/ for wavelength selectivity
10. Filters/wavelength filters such as laser line filters
11. Photodiode/ for power feedback and control

- Specifically, alignment of the resonator cavity is confirmed when the beam reflected by the second mirror (HR or OC) of the cavity retroflects the alignment beam to the alignment laser

6.2.3.d. 2. Good Alignment Practices

- The alignment/fabrication technician should be cognizant of the following to minimize frustrations.
    1. <u>Zero Degree Incidence:</u> Generally, beams must be incident onto optics at zero degrees, with respect to the normal, or else this may lead to beam degradation or aberrations such as astigmatism
    2. <u>Centeredness:</u> In general, laser beams must be centered otherwise off optic axis/center beams will lead to misalignments, stray beams, coma, beam clipping etc.. Centeredness is particularly critical when the optical surfaces are not flat i.e. curved.
        a. By default, optics surfaces must be intercepted by laser beams at zero angle of incidence otherwise the beam be distorted.
        b. If angular orientation is special for best performance mount at the correct angle e.g. beam at Bragg's and Brewster's angle.

Figure 6.5. Uncentered, scattering and clipping alignment beam

Figure 6.6. An active-medium centered alignment beam

3. <u>Beam Clipping:</u> The highest output power and beam symmetry are achieved if the beam is not clipping component edges, e.g. aperture edges.
4. <u>Component Heights</u>- The beam path should be parallel to the work bench
5. <u>Beam Overlap at Transparent Objects:</u> When aligning through transparent components, e.g. Brewster's window/plate/block, the beam reflections at both interfaces (front and back surfaces) should overlap as verification that the component is aligned.

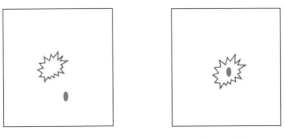

Misaligned beam   aligned beam

Figure 6.7. Confirming alignment of retro-reflected laser beams.

For example, in the case of a Brewster block if the incident beam is reflected off the air-window interface to form a dot while the smudge is a reflection off the window-air interface, when the dot and smudge overlap that confirms that the beam is now aligned at both block interfaces.

6. Torque Settings: Always use the recommended torque settings for screws when mounting components.
    a. Your screws can not be too tight or too loose.
        i. In either case, this would create mechanical stress/imbalance
    b. If too tight you could break-off the screw heads and /or damage and weaken threads.
    c. If too loose they could come off easily during a mechanical vibration thus leading to misalignment.
        Also, note that loose screws could affect your laser beam stability tests
7. When aligning components note the following
    a. Align/Tweak HR and OC, and all component in-between if necessary, to get maximum power out of the active medium.
    b. Orientation of coated surfaces in relation to the laser beam
        - For example, the coating on mirror surfaces in relation to the active medium
    a. It is prudent to slightly misalign some components, such as q-switches to avoid "intracavity lasing" or the "etalon effect'.

6.2.4 To Lase or Not to Lase

- Do not turn on the AC power supply until after all electrical connections in the laser have been made and checked.
- *A laser should never be operated with the cooling system off unless ambient temperature is deemed enough*
    - Turn on cooling device several minutes before turning on and after turning off laser.
    - Always set the cooling device to the correct temperature.
        - Do not turn on laser until the temperature has stabilized at the correct temperature.

Figure 6.8. Temperature indicator on a chiller.

- If a chiller or heat exchanger is used for cooling and a specific flow rate is recommend ensure that as well

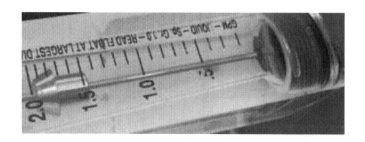

Figure 6.9. Flow-rate indicator of a closed-loop chiller

- Make sure the chiller/heat exchanger has enough coolant in it.

Table 6.1. Common laser cooling devices

	Power Levels	Cooling Methods
1.	Low	Ambient air
2.	Medium	Forced air/Fans
3.	High	Chillers and Heat Exchangers
4.	Up to high power	Thermal Electric Cooling (TEC) a.k.a. Peltier Effect

Note that TEC can be slow effecting temperature changes particularly with high-power lasers (~ 3W). Chillers and heat exchangers have a more rapid thermal response rate/time compared to TEC.

- If your laser has more than one control system e.g. a software-based user interface and a hand-held remote control, turn on only one of them at a time. Having both on will cause communication errors
- After turning on the laser.
  o It should lase/output as soon as the power supply (pump) is turned on if aligned correctly.
    - Safety consideration - If your power supply chassis has a voltage on it it means that there is a short and you are at risk of being shocked.

- o If the laser wavelength is invisible use an invisible beam viewing device, such as an IR viewer, to verify that your laser is lasing.
  - Place a power detector in from of the laser exit port.
    - If there is no reading that means that the laser is not aligned correctly.

6.2.5 Laser Resonant Cavity Output Optimization

The output power optimization process revolves around cleaning and realigning of optics, and temperature scans. All this tweaking must happen in the Current, not Power, Mode otherwise laser settings will be compromised.
So, if output power is less than expected
Look for and
- ✓ clean dirty components
  - o clean optics from the laser exit aperture going back to the HR until the dirty is gone.
  - o remove damaged components
    - Can use heat gun to unglue some components
    - Blow compressed air/nitrogen to clean off and residue
- ✓ Remove any damaged optics because cracks on optics can get worse due to mechanical stress caused by temperature extremes (hold to cold and vice versa).
- ✓ Realign the laser cavity in Current Mode
  - o Align at a low temperature to ensure that the beam is good even at high temperatures.
    - If you align at high temperatures the beam is likely to degrade at low temperatures.
  - o While aligning/tuning the components always seek the global maximum power
- ✓ The following can be adjusted for best laser performance/power
  - o Pump current

- Make sure the pump is getting sufficient excitation energy as per specifications.
- If laser output power goes down when you increment the pump current, it's also an indication of poor alignment or optical pump wavelength change if it's a solid-state laser.
  - Diode temperature – if laser is DPSS
    - The wrong temperature will cause the laser diode to output the wrong pump wavelength and lower output power.
  - Cavity temperature
    - Always temp-scan for a new peak after making some adjustments/changes to the laser components.

If you get max power but unacceptable noise levels, you will have to rescan cavity until you-find a max with acceptable noise levels.

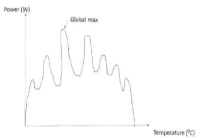

Figure 6.10. Finding the correct temperature for
your laser through cavity temp-scans

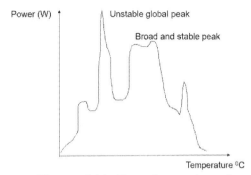

Figure 6.11. Tune in to a broad stable peak and avoid narrow unstable peaks

- If set at an unstable peak point, the laser will quickly slide into the noisy low power valleys.
  - If temperature sensor, e.g. thermistor, electrical connection(s) is broken temperature will not be controllable.
    - Make sure that the thermistor is waterproof
    - Check the resistance of your thermistor to make sure it's within specifications
    - Be ESD-proof when handling thermistors and any electronic components/devices.
  - Also, if the controller is malfunctioning temperature will not be controlled or controlled properly.
    - Malfunctioning controller could cause temperature changes away from optimization
    - Always test controller performance specifications to make sure that it is not malfunctioning

- - a. If applicable, optimize the harmonics crystal phase-matching process being employed for power optimization.
  - All crystal and optics must be clean to get the best output.
    - Harmonic crystals like BBO and LBOs must be temp-tuned.
      - Caution: They can become electrically charged when being heated thus will attract dust if you try to blow clean them while the oven or heating system is on.
    - KTP must be in the correct orientation/angle in relation to the beam for max power.
      - Wrong KTP angle can affect output wavelength and the beam's noise level.
  - If your goal is only to attain maximum power, not beam quality, apertures in the cavity can be widened, or even removed

  - If beam quality is a priority and particularly M^2, insertion or narrowing of apertures in the laser beam will be necessary.

  - Run the laser in current mode until you meet all the performance set-points/performance specifications. Once specs are met the laser can be switched to power/light mode.

Sealing the Laser Cavity
- Follow your employer's directions on how to seal the laser cavity/head
  - Generally, it is required that the head be purged using dry/de-ionized nitrogen, or an equivalent product, as a final step for laser building
- Your laser will have to go Burn-in and Final Test before being released to a user/customer.
- Always seal the exit port of the laser after using it to avoid the inflow of dust and moisture.

- Generally, as housing is sealed the power of the laser drops, so it may be prudent to pre-emptively consider ways to get it back to customer specification. For example, pump current could be raised to compensate for the lost power.
- Always turn off the power source before unplugging the laser, or any systems

Laser Output Performance/Final Tests

You may be required by your employer to test for the following laser performance parameters.

- **CW Output Specifications**
- 1. Power
- 2. Power Stability
- 3. Wavelength
- 4. Beam Width and Ellipticity
- 5. Beam Divergence
- 6. M^2
- 7. Gaussian Fit
- 8. Beam Pointing Stability
- 9. Polarization Ratio and Extinction Angle
- 10. Peak-to-peak Noise
- 11. RMS Noise
- **Pulsed Output Specifications**
- *12. Period*
- *13. Rep Rate*
- *14. Average Power*
- *15. Energy Per Pulse*
- *16. Pulse Duration*
- *17. Duty Cycle*
- *18. Peak Power*

Your laser will have to pass all the required tests before going to the Next step in the production process.

7. Lab 1 Laser Resonant Cavity Fabrication

STEP A: X-Y Alignment System Set-up

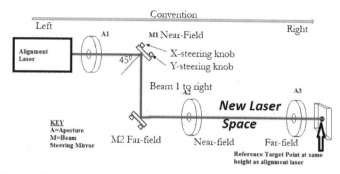

Figure 7.1. laser beam steering apparatus

A1. Put on your alignment laser eyewear and set-up.
A2. Center beam through A1, A2 and A3 to reference target point.
A3. Move A3 toward A2, and A2 toward A3 and adjust M1 and M2 until there is no beam clipping
A4. Turn off alignment laser

STEP B. Align the Active Medium

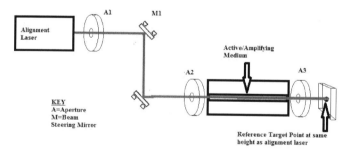

Figure 7.2. Aligning the active medium.

B1. Install the active medium between A2 and A3 as shown in the figure above.

B2. Turn on alignment laser and "walk the beam" through the active medium
   all the way to the reference target point.

STEP C. Install and Align OC

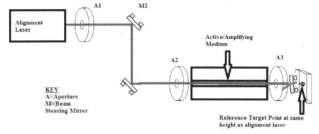

Figure 7.3. Aligning the active medium and OC

Figure 7.4. Picture of a breadboard laser mirror

STEP C2. Align OC Beam Back to Source

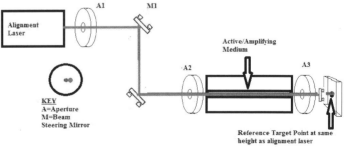

Figure 7.5. Aligning OC with alignment beam

Apertures and Back-reflections as Diagnostics

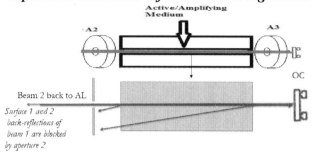

- To make sure you have the correct beam out through A2 wave a semitransparent piece of paper, such as lens cleaning tissue, between the active medium and OC.
- Watch the retro-reflected beam "blink" past A2

Figure 7.6. Alignment multireflection.

STEP D. Install HR and Align Beam to OC

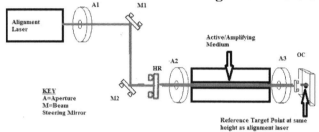

Figure 7.7. Alignment of HR to OC

This task is accomplished by aligning the beam from the OC back to the OC by the HR. However, the presence of multiple reflections can lead to alignment challenges.

The alternative is the align the alignment laser beam off the back of the HR to the alignment laser but without entering it as we did in Step C2. This alignment indirectly aligns the beam from the HR, through the active medium, to the OC. So, we would expect two beam dots near the alignment laser exit port, one from the OC and the other from the HR.

STEP E. Getting Ready for the First Laser

E1. Replace the alignment target with an appropriate power meter.
E2. Turn off the alignment laser.
E3. Remove the alignment apertures.
E4. If the expected laser output is invisible, make sure to get an appropriate viewer. For example, if IR make sure to get an IR Card or Viewer.
E5. Put on the eyewear appropriate for the laser which you have just built.

STEP G. Getting Your First Lase

- If properly aligned a laser should lase after the pump source is turned

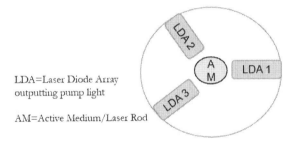

LDA=Laser Diode Array outputting pump light

AM=Active Medium/Laser Rod

Figure 7.8. Cross-section of active medium and pump housing for a side-pumped solid-ststae laser.

- Pump the Active Medium
    - Turn on the pump/power and slowly increase the pump power/energy to energize the active medium until you get a reading on the power meter
        - Beam must be at the center of the power meter detector

Pre-Lase Scenario

1. As you increase pump energy first you get spontaneous emission and eventually population inversion is reached. *The glowing of the active medium may manifest this stage.*

2. Light from the active medium will head in just about every direction and some of it will hit HR and/or OC

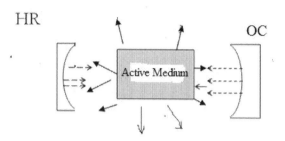

Figure 7.9. Active medium emissions and mirror reflections.

- Threshold is the pump point at which you get your first lase. Make sure to document this point in your laser user's manual e.g. 12 Amps, 20 Watts etc...

- You can increase the pump power/energy up to a desired output power level past threshold.
    - If the laser discharges excessive heat, e.g. creates temperatures above ambient, with ventilation would be appropriate.

Multicomponent and folded resonators.

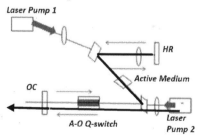

Figure 7.10. A multicomponent z-cavity.

8. Lab 2. Laser Pump Output Wavelength and Tuning

Pre-requisite: You must complete the "Getting Started with the S-P BL-X Laser" procedure to engage in this lab.

DIODE LASER WAVELENGTH TUNING

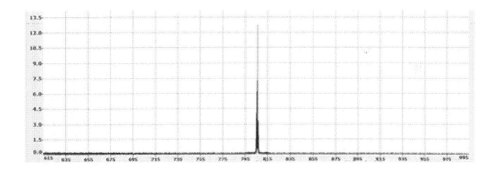

A. OBJECTIVE

The objective of this lab exercise is investigating how a laser diode's output wavelength is affected by both temperature and current.

B. BACKGROUND

Generally, laser diodes are used on their own or as pumps for solid-state lasers.

The temperature of a laser diode can be changed by varying either the (a) temperature of its housing and/or (b) its drive current.

When used as a pump for a solid-state laser, it is critical that the laser diode pump wavelength remains within the absorption band of solid-state laser gain crystal being pumped. It is therefore necessary to determine the laser diode drive current and ambient/housing temperature needed to output the target pump wavelength. In this experiment the target laser diode output wavelength is 808 nm

C. EQUIPMENT NEEDED
- Laser Diode with associated driver software
- OD filters and any attenuators
- Optical Spectrum Analyzer (OSA) or spectrometer with associated computer
- Power Meter and Detector
- Laser eyewear for laser diode
- Optics inspection and cleaning kit(s)

D. REFERENCE DOCUMENTATION
1. Getting Started with the Thorlabs OSA
2. Getting Started with the Spectra-Physics BL Laser

EXPERIMENT

3. Turn the laser.

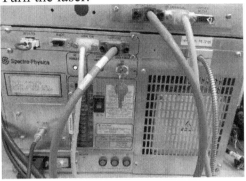

After booting the power supply screen will read "Boot complete" as shown above.

4. Ensure that the laser output power is less than 10 mW, preferably 5mW, to avoid damaging the OSA.

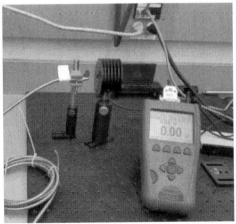

   a. If necessary, use attenuators such as OD filters to reduce the laser diode output to protect the OSA

5. Align diode laser beam into OSA-coupled fiber optic cable

Note:
- If using the ThorLabs OSA click on "Peak Track" to get the peak wavelength after changing the variable parameter, e.g. temperature/current, value and when thermal equilibrium has been reached.
- Always wait for the laser diode to reach thermal equilibrium before recording its output wavelength.

Part 1: Constant Current and Changing Temperature
A. Set the laser current to 4.0, 6.0, 8.0 and 10.0 Amps and at each of these currents change the laser temperature from 15, 20, 25 and then 30 degrees Celsius.

Table 8.1. Sample constant current data table.

Samples	Current (A)	Wavelength (nm) at 15 °C	Wavelength (nm) at 20 °C	Wavelength (nm) at 25 °C	Wavelength (nm) at 30 °C
1	4.0				
2	6.0				
3	8.0				
4	10.0				

Part II Constant Temperature and Changing Current
A. Set the laser temperature 15, 20, 25, 30 and 35 degrees Celsius and at each of these temperatures change the current from 4.0, to 10.0 Amps in 2.0 Amp increments.

Table 8.2. Sample constant temperature data table

Samples	Temp (°C)	Wavelength (nm) at 4 Amp	Wavelength (nm) at 6 Amps	Wavelength (nm) at 8 Amps	Wavelength (nm) at 10 amps
1	15				
2	20				
3	25				
4	30				
5	35				

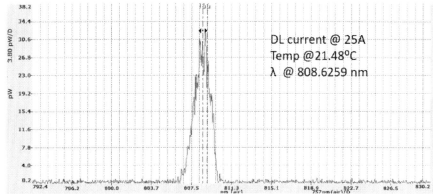
Figure 8.1. Dependence of wavelength on laser diode I

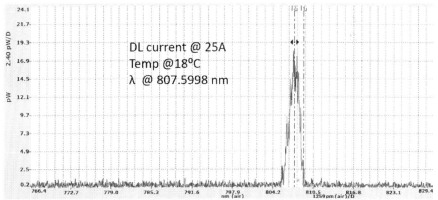
Figure 8.2. Dependence of wavelength on laser diode II

9. Threshold, Slope Efficiency, and Warm-up Time

- After turning a laser on it consumes input or pump energy but does not output
    - This phase is called small signal gain or unsaturated gain
        - In small signal gain losses are greater than gain
- Threshold is when a laser starts to output
    - Threshold kicks in when gain is equal to losses

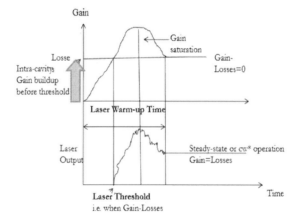

Figure 9.1. Graphical representation of laser energy dynamics after a laser has been turned on.

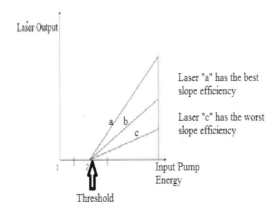

Figure 9.2. Characterization of threshold and slope efficiency.

- **Importance/Meaning**
    o <u>Warm-up time</u> includes small signal gain and saturation time periods
    o It is important to know your laser's warmup time to ensure consistence in performance
    o The <u>threshold</u> point gives an indication of how much energy is consumed before any laser output.
    o <u>Slope efficiency</u> reveals the fraction of energy input that emerges as laser output above/past threshold
- **Issue(s)**
    o Operating a laser in the warm-phase could result in non-uniform or inconsistence performance
    o The shorter the threshold the better.
        - Energy consumed before threshold is "wasted" energy

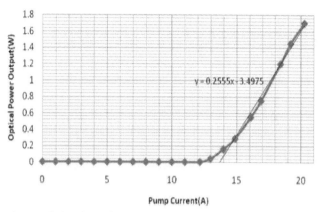

Figure 9.3. Graph of a laser's threshold and slope efficiency.

- In the example, above, a slope efficient of .2556 means that .2556 Watts were output by the laser for every Amp of pump energy consumed
- When both pump input and output power are expressed in the same units, e.g. Power (Watts), it is apparent that a it is impossible to get a <u>slope efficiency</u> of 100%.
    - A slope efficiency of 100% means that for everyone Watt of input 1 Watt of out power was generated.
    - It is therefore recommended that slope efficiency data be requested or presented in the same units for straightforward cost analysis.

Personal Notes

10. Lab 3. Collimation of Diverging Laser Beams

Objective: To collimate a diverging laser beam

Background: Laser beams tend to diverge after they leave the lasing cavity, i.e. the diameter of the beam increases with distance if they are not collimated.

Such divergence can be negated if the beam is simultaneously expanded and collimated i.e. all its individual rays are made parallel to each other and the beam diameter is constant over distance.

According to Diffraction Theory the angle of diverge, θ, of a beam is given by

$$\theta = \frac{k\lambda}{d}$$

Where k is a constant
λ is the beam's wavelength and a constant value as well.
d is the beam size

Examination of this equation shows that as the beam size, d, is increased is divergence will decrease. Also, note that a collimated laser beam will exhibit the smallest spot size/focal point when collimated. Therefore, laser beam collimators are inherently beam expanders and the resulting/exiting beams have lower beam divergences compared to the entrance/incoming beam. In addition, the resulting/exiting beams have the smallest spot size when focused.

Collimation is achieved by using two lenses positioned so that their respective focal lengths coincide at a single or common point, i.e. if they were to focus already collimated beams coming from opposite directions. The collimator lens separation, CLS, would calculated as follows

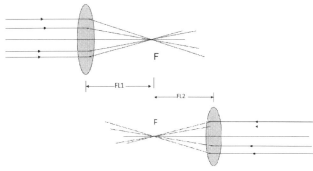

Figure 10.1. Demonstrating how two lenses can have a common focal point.

$CLS = FL_1 + FL_2$

if the incoming beams above were both collimated.

However, a laser beam emanating from a resonator is not generally already collimated therefore the telescopic assumption above is only a good starting point and the technician/experimenter would have to "tweak" the two lenses in the collimator until their two focal points overlap as shown below. In some literature laser beam collimators are referred to as telescopes for reasons stated above

There are two types laser beam collimator/expanders namely Keplerian and Galilean. A Keplerian collimator uses two positives lenses, while a Galilean used one negative and one positive lens as shown in Figure 3. A Keplerian Collimator/Expander is more suitable for low power beams because high energy/power densities are produced at beam waist that could cause thermal issues such as the ionization of air inside the collimator etc..

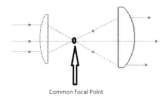

Figure 10.2. Plano-convex Keplerian telescope/collimation with the two plano sides of the lenses "sandwiching" the common focal point.

If plano-convex lenses are used the convex sides should be facing away, or farthest, from the common focal, while the focal point is "sandwiched by the plano sides of both lenses as shown in the Figure 3.

The laser beam Galilean Collimator/Expander is more appropriate for high power/energy beams since it reduces beam irradiance/fluence.

Keplerian Laser Beam Collimator/Expander

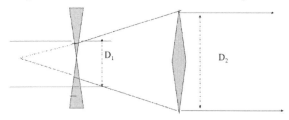

Figure 10.3. Galilean Laser Beam Collimator/Expander

It is common industry practice to insert a laser beam collimator/expander is at the exit port of a laser resonator after the laser has been fabricated. The collimator/telescope lens separation distances is optimized before sealing and shipping the laser to a customer/user.

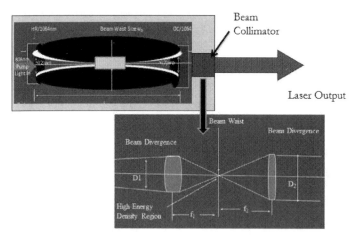

Figure 10.4. A collimator/telescope at the beam exit of a laser allows for controllable adjustments to beam quality parameters such as M-squared and divergence.

We shall construct a Keplerian laser beam collimator/expander in this lab exercise.

We can use at least one of the following to test how well our laser beam collimator/expander works.
a. beam size after collimation
b. beam spot size after focusing the collimated beam.
c. angle of divergence
d. M^2 value

A Share Plate Collimation Tester lens can also use as shown below

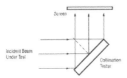

Use Plate at 45 Degree AOI for Best Results

The edge of the optic is marked with an arrow indicating the thickest part of the optic. The arrow points toward the front surface of the optic, which should face the incident beam. For a horizontally propagating beam the mark should be positioned on top so that the wedge is perpendicular to the plane of incidence.

Interpreting Shear Plate Collimator Tester Fringes

When horizontal beam is collimated, fringes will be horizontally oriented. Fringe orientation will be tilted slightly when beam is diverging or converging.

Figure 10.5. Source: https://www.newport.com/f/shear-plate-collimation-tester

In this lab you will have to measure the collimation indicator values before collimation and compare with the after-collimation values.

EQUIPMENT NEEDED
1. HeNe Laser /any low power laser
2. Rail at least 30 cm long
3. 3 post holders to mount on rail
4. 3 posts
5. 3 lens mounts/holders
6. 3 plano-convex lenses of Focal Length (FL)

7. Graduated Ruler (3 ft or 1 m)
8. Beam Profiler

Procedure:

Determine approximate focal length of Lens:
The focal length (FL) of the convex lenses issued to you needs to be determined or verified. Use methods learned in Laser 100 to accomplish this task.

Table 10.1. Lenses' Focal Lengths

	Lens	Focal length (mm)
FL1	Lens 1 (shortest focal length)	
FL2	Lens 2 longest focal length)	
FL3	Lens 3 (Intermediate focal length)	
FLsum	Sum of focal Lengths	

As a rough guideline, the sum of the focal lengths of the lenses cannot be greater than the length of your optical track or rail system. Make the track length be greater than the sum (f1+f2+f3) by at least 10cm.

A. Rough Collimation Beam Alignment:

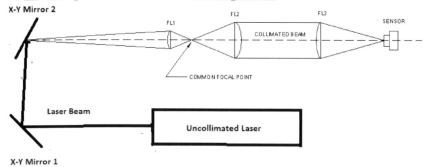

Figure 10.6. Laser beam collimation set-up.

1. Mount a rail track on an optical bench that is at least 1m long
2. Mount a profiler sensor/detector at the end of the rail
3. Turn on the profiler
4. Mount the laser in an orientation whereby you can use the two x-y mirrors to walk the beam to the sensor
5. Place mirror 2 in a position to reflect the laser beam onto optical rail or track.
6. Have the laser, mirrors and profiler sensors at the same height. Specifically, their centers have to be at the same height.
7. "Walk the beam" from the laser into the profiler detector. View the screen of the profiler to make sure that it registers the beam position. Make a note of this beam position for this step helps you define your optical axis from the laser to the profiler.

<u>Before you mount the lenses (steps 8 and 9) note that</u> plano-convex lens must be oriented so that the flat side is toward the focal point of a beam

8. Mount and position lens 1 (the one with the shortest focal length (f_1)) on the opposite end (from the profiler) of the track. Since everything was well aligned in the previous step(s), only "tweak" this lens until the beam is back onto the optical axis. Check your profiler to verify this. ***To be successful, aim to pass the beam through the optical axis, or center of the lens***.
9. Move the profiler sensor toward lens1 until it registered the smallest spot size. Mark that point on the rail. *This is necessary particularly if the incoming beam is not already collimated which means that CLS will not be equal to FL1+FL2.*
10. Move the profiler sensor back a distance greater than FL2
11. Mount and position lens 2 (the one with the longest focal length (FL_2)) a distance equal to the FL2 from the spot marked in the step 9 above.
12. Again, since everything was well aligned in the previous step(s), only" tweak" the lens 2 until the beam is back onto the optical axis. Check your profiler to verify this. ***To be***

successful, aim to pass the beam through the optical axis, or center of the lens.
13. If the distance from the FL1 focal point to lens 2 is equal to FL2 the beam should be collimated .
    a. To verify the collimation, move the profiler sensor back and forth the rail to see if the beam maintains the same size everywhere along this path. If OK **the beam is now roughly collimated**!!!
14. Call the instructor before moving on to Fine Collimation Alignment.

X_____

B. Collimation Refinement.
B-1 Beam Size as a Measure of Collimation
1. Mount the 3rd lens i.e. Collimation Test Lens, of intermediate focal length, between lens 1 and lens 2.
2. Align the beam until its back on the optical axis after inserting lens 3.
3. Place the profiler sensor at a distance exactly equal to its focal length (FL3 or f_3).

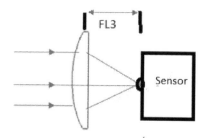

Figure 10.7. Collimation Test lens must be placed a distance equal to its own focal length, f_3, from the profiler beam sensor.

4. Move lens 1 only until the beam registers the smallest beam spot size on the profiler screen

a. Note that a collimated beam will have a beam waist i.e. smallest beam size at its focal length i.e. at the location of the sensor as shown in Figure 8.

Call Instructor

X_____

B-2 Beam Divergence as a Measure of Collimation

a. For the collimated beam measure D_{FL}, which should also be D_{min} if the beam is truly collimated
$$f_3\theta = D_{FL} \qquad (\text{Recall } FL_3 = f_3)$$

So, in this case
$$\theta_{col} = D_{FL3}/FL_3$$

a. Remove the collimator i.e. lenses 1 and 2 and compute the angle of divergence of the raw beam. For the un-collimated beam measure D_{FL3} will not coincide with D_{min} so make sure to keep the profiler sensor at FL_3 (=f_3).

$$\theta_{raw} = D_{FL3}/FL_3$$

Was the beam divergence greater with or without the collimated and expanded beam? Why?

B-3. M^2 as a Measure of Collimation

1. Measure the M-squared value of the beam without the collimator
2. Measure the M-squared value of the beam with the collimator

Table 10.2. How M-squared is affected by collimation.

	M-squared
No Collimator	
With Collimator	
% Difference	

B-4. Compute Telescopic Magnification/Beam Expansion

Experimental Magnification
Measure the beam size of the collimated and expanded beam
$D_2=$
Measure the raw beam size
$D_1=$
Compute the magnification, M
$M_{exp}=D_2/D_1=$

Theoretical Magnification

$D_2=D_1|(FL_2/FL_1)|$
so
$M_{theo}=D_2/D_1=FL_2/FL_1$

Comparison of magnifications

% Diff in Magnification $= \dfrac{|M_{exp}-M_{theo}|}{M_{ave}} \times 100$

11. Lab 4. Laser Threshold and Slope Efficiency

Pre-requisite: You must have completed at least two solid state start-up procedures to engage in this lab exercise.

Warnings:
1. Chiller must be running before laser is fired, where applicable.
2. Appropriate laser eyewear must be worn before the laser is fired.
3. Power must be attenuated to zero then slowly remove ND filters until a signal is observed.

Make sure to know the operation and limitations of the laser analyzer you are using to avoid damaging it and getting the wrong measurements

Objective:
To determine the threshold and slope efficiency of a various lasers.

Materials Needed:
A. Laser eyewear
B. 1 Power meter
C. At least one of the following lasers but preferably all
   a. end-pumped solid-state laser
   b. diode-pumped solid-state laser
   c. lamp-pumped laser
   d. (electrical-pumped) gas laser
   e. (electrical-pumped) diode/semiconductor laser

Experiment:
Refer to the "Getting Started" procedure you developed for each laser above.
1. Vary the input pump energy/current in unit increments from zero/minimum until the laser reaches threshold. Plan on taking equally 3 pre-threshold data points.
2. After threshold wait until the laser goes past saturation before increasing the input energy/current. Refer to your "Getting Started" procedure for the specific laser saturation phase time.

3. After saturation, you would be in the steady-state region, take data for slope efficiency.
    a. Plan on taking at least 7 data points/sets past saturation until output power maxes out.

Data:
1. For each laser record input pump energy (x) and laser output power (y) in a Table.

Table 11.1 Sample Data Table

	Pump Input (Units)	Laser Output (Units)
1		
2		
3		
4		
5		
6		
7		
8		
9		
10		

Where applicable Chiller must run for at least 15 minutes after the laser is turned off to ensure that the laser rod does not get damaged.

Computations:
For each of the lasers compute as instructed below

Plot/graph Pump Energy (x) vs. Power Output (y) using Excel and glue its hardcopy in your note book.
    a. Use Excel's Trend-line function
        "Eyeball-metrics" of power meters and graphs are good practices to quickly estimate threshold energy points of lasers etc. However, the most reliable method is to utilize a regression models such as accorded by MS Excel Trends. Using such mathematical models, you can also

determine the slope efficiency, in addition to the threshold energy point.

Plot your data for each of the lasers under study and extract a regression/trend-line equation. Using this equation determine
  i. Threshold and slope efficiency

Results Reporting:
Prepare a poster and oral presentation to present your findings.

Additional Experimental Notes on Threshold and Slope Efficiency

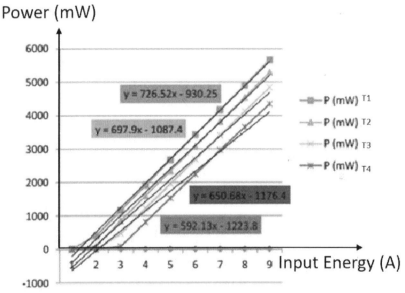

Figure 11.1. Dependence of Threshold and Slope Efficiency on Temperature. T4>T3>T2>T.

12. Lab 5 Laser Beam Harmonics Conversion Efficiency

5a. Objective: The objective is to demonstrate laser beam second harmonic generation conversion efficiency is affected by interceding optics and their cleanliness.

Equipment Needed:

1. 2 lasers of different wavelengths to supply the fundamental beam (P_f)
2. A SHG Box or such set-up
3. Power meter
4. Spectrum analyzer/spectrometer

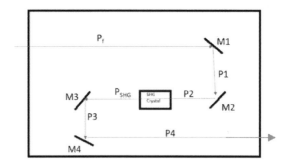

Figure 12.1. The SHG Box set-up.

Experiment: Turn of the fundamental laser and measure P_f, P1, P2, P_{SHG}, P3, and P4 using a power meter. Measure the output wavelength, λ_{SHG}.

Data and Computations

Laser 1
Table 12.1. Laser Powers and Efficiencies

		P(mW)		Efficiency (%)
1	P_f			
1	P1		$P1/P_f$	
2	P2		P2/P1	
3	P_{SHG}		$P_{SHG}/P2$	
4	P3		$P3/P_{SGH}$	
5	P4		P4/P3	
6		CE*	$P4/P_f$	

*CE=Conversion Efficiency

Table 12.2. Fundamental and SHG Wavelength Relationship for Laser 1

		λ (nm)	
1	λ_{1f}		
2	λ_{1SHG}		
3	λ_1 Ratio	$\lambda_{1SHG}/\lambda_{1f}$	

Table 12.3. Laser 2 efficiency data

		P(mW)		Efficiency (%)
1	P_f			
1	P1		$P1/P_f$	
2	P2		P2/P1	
3	P_{SHG}		$P_{SHG}/P2$	
4	P3		$P3/P_{SGH}$	
5	P4		P4/P3	
6		CE*	P4/Pf	

*CE=Conversion Efficiency

Table 12.4. Fundamental and SHG Wavelength Relationship for Laser 2

		λ (nm)	
1	λ_{2f}		
2	λ_{2SHG}		
3	λ_2 Ratio	$\lambda_{2SHG}/\lambda_{2f}$	

Clean the optics and repeat the experiment for power efficiencies only.

Table 12.5 Laser 1 efficiency data

		P(mW)		Efficiency (%)
1	P_f			
1	P1		P1/P_f	
2	P2		P2/P1	
3	P_{SHG}		P_{SHG}/P2	
4	P3		P3/P_{SGH}	
5	P4		P4/P3	
6		CE*	P4/Pf	

*CE=Conversion Efficiency

Table 12.6. Laser 2 efficiency data

		P(mW)		Efficiency (%)
1	P_f			
1	P1		P1/P_f	
2	P2		P2/P1	
3	P_{SHG}		P_{SHG}/P2	
4	P3		P3/P_{SGH}	
5	P4		P4/P3	
6		CE*	P4/Pf	

*CE=Conversion Efficiency

13. Electrical Outlets, Power Supplies, and Wall Plug Efficiency

- This is important particularly when the equipment does not draw electricity from conventional electrical outlets.
- For example, in North America if the laser/laser system, or its support equipment, are run at other voltages, other than conventional 120 Volts, it may be necessary to have those special electrical outlets installed if they are not already in place and at the desired location.
- If you do not plan, by first asking your sales rep or carefully reading the laser/laser system electrical specification sheet, you may face delays in using your new laser/laser system since you will need to have the special outlets installed in your facility.
- Importance
    - Electrical signals drawn by power supplies from electrical AC outlets can be of any AC Voltage
    - It is important that the <u>electrical outlet</u>/ power supply input requirements for a laser/laser system be specified/inquired to ensure it would fit with existing electrical outlets in the hosting building/room
    - In addition, electrical energy extracted from power lines by lasers undergoes various transformations before it exits a laser as light energy. During the transformations, some of the energy is lost leading to inefficiency. Some lasers lose less energy than others thus making them more attractive from a cost point of view
- Issues
    - If the <u>electrical outlets</u> of a purchased laser/laser system do not match your existing power outlets additional costs will be incurred to rewire outlets
        - In addition, you may have to consult with a power electrical engineer to ensure that the laser power supply will be sending the correct voltage(s) to the laser etc.
    - Lasers power supplies often draw AC electricity first then convert it to DC.

- This conversion efficiency is never 100% thus leads to AC noise that is carried over to the laser
    - Laser performance parameters include root-mean-square (rms) and peak-to-peak noise which sometimes can be traced to the power supply
    - Your laser output, at best, will have the same noise as your power supply, however, more often it would be higher. It is therefore prudent that the power supply used outputs less noise than the acceptable laser beam output noise.

<u>Wall-plug Efficiency</u> is the ratio of output optical power to input electrical power thus gives a comprehensive sense of the energy cost of running a laser/laser system

14. Lab 6. Laser Power Supply and Wall-plug Efficiency

Objective: The objective of this lab is to determine the wall-plug efficiency of three lasers [solid state, ion and neutral gas]

Introduction: Electrical energy extracted from power lines by lasers undergoes various transformations before it exits a laser as light energy. During the transformations, some of the energy is lost leading to inefficiency. Some lasers lose less energy than others thus making them more attractive from a cost point of view.

Equipment:
- A. Laser eyewear for all team members
- B. One Power meter
- C. One RMS clamp-on amp meter
- D. At least two of the following lasers
    - a. end-pumped solid-state laser
    - b. diode-pumped solid-state laser
    - c. lamp-pumped laser
    - d. (electrical-pumped) gas laser
    - e. (electrical-pumped) diode/semiconductor laser

Warning(s)

Observe all electrical and laser safety guidelines.

Experiment:

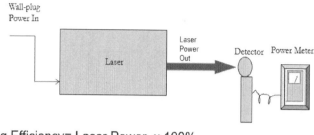

Wall Plug Efficiency= $\dfrac{\text{Laser Power}}{\text{Wall Power}} \times 100\%$

Figure 14.1. Experimental set-up for wall-plug efficiency.

1. Plug the laser into the clamp on meter adapter and the adapter into an outlet

2. Clamp the meter around the adapter. Turn on meter measure AC amps.
3. Aim the laser into the power meter.

Procedure: use procedures developed in the past to operate the lasers.

Data:
For each laser under study make a table in your lab-notebook to record current/energy being extracted from the wall-plug by the power supply (PS) and associated laser output power.
Divide the input energy/current into at least 10 data extraction points.

Sample Data Table

Laser 1 Name: _____

Set Temperature (If applicable): _____

Table 14.1. Sample Data Table for Laser 1

	PS Input Current (A)	Laser Output Power (W)
1		
2		
3		
4		
4		
5		
6		
7		
8		
9		
10		

Laser 2 Name: _____

Set Temperature (If applicable): _____

Table 14.2. Sample Data Table for Laser 2

	PS Input Current (A)	Laser Output Power (W)
1		
2		
3		
4		
4		
5		
6		
7		
8		
9		
10		

Computations:
1. Compute the wall-plug efficiency for each laser and compare.

15. Relative Stability Fluctuations and ISO Standard of Stability Tests (ISO 11554)

- The output of all lasers fluctuates, and perhaps worsen over time
- Generally, on cw power stability fluctuations are specified during a transaction
    - The other parameters' stability fluctuations can be provided by the sellers/manufacturers if requested.

ISO Standard of Stability Tests (ISO 11554)
- Generally, parameter stability fluctuations exhibit a Normal Distribution

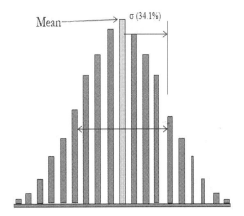

Figure 14.2. Normal Distribution Plot

- In a normal distribution measurement have a symmetric distribution around the mean
- Standard Deviation, σ, is a measure of the data spread, or deviation, from the mean.
- About 68.2 % of the measurements ,2σ , are clustered around the mean
- The larger the standard deviation the less precise or repeatable a measurement is.

- ISO Standard of Stability Tests (ISO 11554).
  The relative fluctuation of a parameter, ΔP,
  $\Delta P = 2\sigma/P_{mean}$
  where σ is the standard deviation and P_{mean} is the mean of the parameter over the measurement period

Laser Parameters' Relative Stability Fluctuations
- Laser beam performance parameter fluctuations general exhibit a Normal/Bell/Gaussian distribution thus this ISO 11554 is applicable and has been used
- Acceptable levels of fluctuations will depend on the application and the users of the laser.
- Relative Stability Fluctuation discloses a specific parameter's performance precision in the laser output

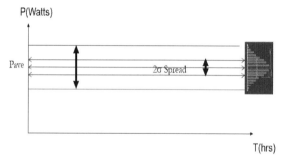

Figure 14.3. Cartoon of a laser stability measurement system

- It is prudent to inquire about the Relative Stability Fluctuations for laser output parameters, or at least those that impact the parameter of interest in an application
  - Moreover, it is prudent for the laser buyer/user to specify *a priori* an acceptable level of relative fluctuation for each of the performance parameters of interest in the purchase agreement/contract

- Key Pulsed Stability Output Parameters
    - Pulse Energy Stability
    - Pulse Width Stability
    - Average Power Stability
    - Average Peak Power Stability
    - Average Rep Rate Stability
    - Etc.
  - *In all cases, it is recommended **that ambient room temperature variations** be specified within which the stabilities hold true.*
    - Typical Specification
      - The relative stability fluctuations can be specified as follows
        - Pulse Width$(\Delta t)_{ave}$ of 80 ns with a stability fluctuation of 3% over 48 hours **between 20 and 40 ^{0}C**

16. Useful Lifetime and Burn-in, and Footprint

- Useful Lifetime and Burn-in
    - The manufacturer usually knows the useful lifetime of a laser from Accelerated Aging Tests. Burn-in tests are used indicators of a laser's useful lifetime.

- Importance/Meaning
    - Knowledge of useful lifetime of a laser/laser system would be valuable to the user in terms of
        - estimating the Return on Investment (ROI)
        - forecasting/planning
        - budgeting
- Issues(s)
    - Laser useful life-time information can made available to the user/buyer but only upon request

Footprint

- Footprint
    - The amount of space a laser/laser system would occupy can be a critical factor but may not be discussed at the transactional stage.
- **Importance/Meaning**
    - Lasers could differ in size by several magnitudes but still deliver comparable performance
- **Issue(s)**
    - If space is scarce or expensive the footprint should be considered
        - Space issues could range from room and/or optical bench sizes
        - In addition, the footprint of support equipment such as chillers, power supplies must be considered as well

17. Laser Fabrication, Factory-level Tuning and Performance Tests Self-Test

Laser Optics Quality and Damage Control Self-Test
(Graded versions of these tests are available online at laserpronet.com. Click on Laser IQ Scan)

1. _____ notation can be used to indicate surface defects
    A. Scratch-Dig
    B. Microscopic notation
    C. A and B
    D. None of the above

2. Optics can be damaged _____.
    A. by laser beams
    B. during cleaning
    C. All the above
    D. None of the above

3. Optical defects include _____.
    A. inclusions
    B. pits/digs
    C. scratches
    D. All the above
    E. None of the above

4. Voids in an optic include _____.
    A. pits/digs
    B. scratches
    C. inclusions
    D. A and B
    E. None of the above

5. Voids _____.
    A. concentrate laser energy within them
    B. can experience dielectric breakdown damage
    C. A and B
    D. None of the above

6. Inclusions/impurities within an optic bulk/inside _____.
    A. can change the oxidation state of the optic site
    B. create absorbing color centers
    C. cause internal damage
    D. all the above
    E. None of the above

7. Optical damage is characterized by defects on the _____ of an optical material
    A. surface
    B. bulk/interior
    C. A and B
    D. None of the above

8. Optical damage can be _____.
    A. Laser-induced
    B. Human- induced
    C. Manufacturing defect
    D. All of the above
    E. None of the above

9. _____ has/have lower damage threshold than the bulk.
    A. Imperfections
    B. Coatings
    C. A and B
    D. None of the above

10. Optical damage can be detected through _____.
    A. visual inspections
    B. performance decline of an optic
    C. A and B
    D. None of the above

11. _____ raise the damage threshold
    A. Dielectric coatings
    B. Super-polishing
    C. A and B
    D. None of the above

12. Super polishing optics could _____ damage threshold.
    A. could lower
    B. could increase
    C. any of the above
    D. has no effect on
    E. None of the above

13. _____ can lower the damage threshold of an optics.
    A. Dielectric coatings
    B. Annealing
    C. A and B
    D. None of the above

14. Defects or imperfections on optics are expected more on _____ optics.
    A. coated
    B. uncoated
    C. A and B
    D. None of the above.

15. _____ on optical surfaces can be "burned-in" when hit by a laser beam.
    A. Solvents
    B. Dirt
    C. A and B
    D. None of the above

16. Which of the following can lower the damage threshold of the surface?
    A. dirt
    B. fingerprints,
    C. All the above
    D. None of the above

17. Optical damage can be lowered by _____ of optics.
    A. excessive cleaning
    B. not cleaning optics
    C. All the above
    D. None of the above

18. Optics can be damaged _____.
    A. By Laser beams
    B. During cleaning
    C. All the above
    D. None of the above

19. Dirt on optics can _____.
    A. cause beam scatter off surfaces
    B. absorb laser energy
    C. Create hot spots
    D. All the above
    E. None of the above

20. Optical damage generally occurs on _____.
    A. perfectly polished optical surfaces.
    B. scratches
    C. digs
    D. coatings
    E. All the above
    F. None of the above

21. Which of the following increase damage threshold of the surface?
    A. airborne particles
    B. fingerprints,
    C. residue from out-gassing of nearby components.
    D. All the above
    E. None of the above

22. _____ will cause damage to optics surfaces.
    A. handling optics with bare hands
    B. fingerprints
    C. finger cots
    D. A and B
    E. None of the above

23. Laser beams must pass through the _____ of optical components for the best beam quality
    A. center
    B. edge
    C. half-way between the center and the edge
    D. any of the above
    E. None of the above

24. Dirty optics can compromise beam _____.
    A. power
    B. energy
    C. quality (M^2)
    D. All the above

25. When optics is being handled _____ must be worn
    A. finger cots
    B. gloves
    C. hair bonnet
    D. all the above
    E. None of the above

26. Excessive optics cleaning would _____ optics
    A. damage
    B. Improve the performance of

27. It is best to _____ clean optics to preserve its integrity and performance.
    A. not to
    B. always

28. _____ will protect optics from being damages
    A. finger cots,
    B. gloves and
    C. tweezers
    D. All the above
    E. None of the above

29. If cleaning solvent containers are not closed/capped the solvent will _____.
    A. pick-up moisture from the air.
    B. degrade and damage optics
    C. A and B
    D. None of the above

30. Optics contaminants include _____.
    A. dust and lint,
    B. fingerprints,
    C. sweat
    D. None of the above

31. Cleaning _____ an optic
    A. decreases the risk of damaging
    B. increases the damage threshold on an optic
    C. all the above
    D. None of the above

32. Cleaning an optic could _____.
    A. cause scratches
    B. implant inclusions
    C. A and B
    D. None of the above

33. _____ benchtop will prevent an optic from getting damaged
    A. edge restraints around a
    B. clean room tissue pad on
    C. placing optics bare
    D. A and B
    E. None of the above

34. _____ will prevent an optic from being contaminated
    A. clean bench top
    B. tweezers
    C. hemostats
    D. all the above
    E. None of the above

35. _____ may alter the specified performance of a given optics.
   A. dirt
   B. damages
   C. A and B
   D. None of the above

36. It is a good laser lab practice _____ when grabbing optics
   A. Use sharp tools to grab optics
   B. Grab optics along the center face(s) through which a laser beam would propagate
   C. All the above
   D. None of the above

37. It is a good laser lab practice to test the resilience of an optic by_____
   A. Scratch the optics coatings with your fingernails
   B. Drop optics onto the optical bench to test their hardness
   C. use tap water to clean the optic
   D. All the above
   E. None of the above

38. It is a good laser lab practice to
   A. keep optics in a clean environment
   B. Wrap optics with appropriate tissue and store away when not in use.
   C. All the above
   D. None of the above

39. It is a good laser lab practice to _____ the first time you plan to use it
   A. clean optics
   B. blow compressed air/nitrogen to clean optics.
   C. Apply solvents such as acetones, methanol, to clean it
   D. Inspect the optics
   E. All the above

40. It is a good laser lab practice to have _____have
    A. A round table while working on optics
    B. A soft tissue pad to lay the optics
    C. wear glove to handle optics
    D. B and C
    E. All the above

41. Wearing gloves ensures that _____.
    A. you avoid contaminating the optics
    B. protect yourself from solvents
    C. A and B
    D. None of the above

42. _____ optics cleaning could damage the optics
    A. Excessive
    B. None
    C. A and B
    D. None of the above

43. To avoid optics damage, it is best _____.
    A. not to clean them if not dirty
    B. clean when necessary
    C. If cleaning is necessary, start with to contact optics cleaning methods first and then resort to non-contact methods last
    D. A and B
    E. None of the above

44. Which of the following can decrease the damage threshold of an optic
    A. Increasing pulse duration.
    B. Irradiating shorter wavelengths
    C. by exposing the component to higher fluence/irradiance levels beginning well below single shot damage threshold.
    D. b and c
    E. None of the above

45. It is good practice to _____
    A. make direct contact with the optics
    B. handle optics from the edges
    C. wipe optics only once or make one pass
    D. B and C
    E. None of the above

Laser Cavity Alignment

The following two questions refer to the figure below

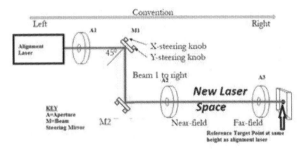

Figure 2. Laser Beam X-Y Control

46. Near-field alignments can be accomplished using mirror number
    A. M1
    B. M2

47. Far-field alignments can be accomplished using mirror number
    A. M1
    B. M2

48. A good rule of thumb while replacing laser and laser system components is to
    A. (1) remove all the bad optics then (2) install the new and (3) align
    B. (1) remove one bad optic at a time (2) install the new and (3) align
    C. (1) remove one bad optic at a time (2) clean the new optic (3) install the new and (4) align
    D. (1) remove one bad optic at a time (2) inspect the new optic and clean it if necessary, (3) install the new and (4) align
    E. None of the above

49. The key reason(s) for aligning a laser resonator is to _____.
    A. give birth to a laser
    B. adjust the power
    C. improve beam quality
    D. All the above
    E. None of the above

50. The key reason (s) for aligning a laser system is to _____.
    A. guide/direct the beam
    B. improve the quality of the beam
    C. All the above
    D. None of the above

51. When aligning a visible beam it is best to use ____ to locate the position of the beam.
    A. IR Viewer
    B. IR Card
    C. The palm of your hand
    D. Lens tissue
    E. All the above None of the above

52. A ND:YAG or ND:YVO₄ laser cavity can be aligned using a
    A. HeNe laser
    B. Low power 1064 nm laser
    C. a and b
    D. None of the above

53. When aligning a laser cavity, the alignment beam enters the laser cavity through one of the mirrors, is "walked" through the cavity components to the _____, then back through the cavity and to the alignment laser.
    A. HR
    B. OC
    C. a or b
    D. iris diaphragm
    E. None of the above

54. If properly aligned a laser should lase as soon as _____.
    A. the pump source is turned on
    B. the cavity/head is purged
    C. chiller
    D. a or b
    E. a and b
    F. None of the above

55. A laser _____ should be turned first before the laser is turned on.
    A. cooling system
    B. washer
    C. lab room lights
    D. All the above
    E. None of the above

56. A laser _____ should be turned after the laser is turned off.
    A. cooling system
    B. washer
    C. lab room lights
    D. All the above
    E. None of the above

57. Aligning a Nd:YAG 1064 nm laser cavity with low power HeNe 632.8nm alignment beam may make it necessary to "fine-tune" the alignment after the laser starts lasing with a _____ laser.
    A. 1064 nm
    B. 632.8 nm
    C. 488 nm
    D. Any of the above

58. It is necessary to get a new laser rod/active medium if the old one _____.
    A. is dirty
    B. has burn marks
    C. is cracked
    D. b and c
    E. all the above
    F. None of the above

59. If a laser's wavelength _____ nm it is invisible.
    A. 532
    B. 633
    C. 515
    D. 1064
    E. None of the above

60. If a laser's wavelength _____ nm it is visible.
    A. 266
    B. 1064
    C. 515
    D. 1064
    E. None of the above

61. The power meter shown is reading about _____ Watts.

    A. 4
    B. 40
    C. .4
    D. All the above
    E. None of the above

After building/repairing a Class 3B and 1064 nm you tested the beam and came-up with the following report.

Table 1: Laser Performance Test Report Sheet

	Cw Performance Parameter Name	Results
a	Wavelength (nm)	1066 nm
b	Power (Watts)	500mW
c	M^2	1.0
d	Beam Divergence	.5mR
e	Beam Roundedness	.6

62. Which parameters can still be improved?
    A. Wavelength, Power, and Beam divergence
    B. M^2
    C. Beam Roundedness
    D. A and c
    E. All the above
    F. None of the above

63. Which parameters cannot be improved or are optimized.
    A. M^2
    B. Beam Divergence
    C. Power
    D. All the above
    E. None of the above

64. Another technician tested this laser right after you but got a Beam Roundedness value of 1.04 and yet you got .6. Why are the values different?
    A. They are not different but mean the same thing.
    B. The beam diameters were swapped when calculating Roundedness.
    C. Laser beam roundedness is very unstable, so this is expected.
    D. a and b
    E. None of the above

65. You spoke to your supervisors about the issue, in 3 above, but tells you that (s)he has never heard of term Beam Roundedness, but you also notice that he is working on a laser performance report with the terms listed below on it. Are any of them equivalent to Beam Roundedness
    A. Beam Ellipticity
    B. Beam Circularity
    C. Beam Divergence
    D. a and b
    E. None of the above

66. Optical alignments can optimize beam _____.
    A. spatial profile
    B. transverse mode(s)
    C. All the above
    D. None of the above

67. Laser power meters should be calibrated
    A. annually.
    B. every month.
    C. every two years.
    D. only when they start to give wrong readings.

68. When calibrating a laser feedback system adjustment must be made such that its internal power meter's reading matches that displayed by a(n) _____ external power meter.
    A. calibrated
    B. grandfathered
    C. uncalibrated
    D. collimated
    E. all the above
    F. None of the above

69. Which analyzer would you use to verify the M^2 value of the laser/laser system
    A. Beam Profiler
    B. Oscilloscope
    C. Power Meter
    D. Spectrum Analyzer/Spectrometer
    E. None of the above

70. Which analyzer would you use to verify that the only wavelength being output is the third harmonic of 1064nm?
    A. Beam Profiler
    B. Oscilloscope
    C. Power Meter
    D. Spectrum Analyzer/Spectrometer
    E. None of the above

71. Optical alignments can correct and optimize_____
    A. beam shape
    B. power
    C. transverse mode
    D. pump current
    E. a, b and c
    F. all the above

72. _____ in a laser cavity will favor higher order transverse modes.

    A. Misalignments,
    B. tilts,
    C. dirt
    D. a and b
    E. All the above
    F. None of the above

73. If your fiber-coupled thirteen Watt diode laser array for pumping your Nd:YAG laser fails to meet this power specification what would be the likely reason(s)
    A. The fiber cable aperture is dirty
    B. one or more laser diodes in the array is failing
    C. the laser head needs to be realigned
    D. the chiller is not turned on

74. Aligning the laser cavity with an alignment beam is to ensure that
    A. the lasing beam hits both HR and OC at right angles to the planes/surfaces of the two mirrors
    B. the lasing beam hits both HR and OC along the cavity optical axis
    C. all contaminants are sanitized
    D. a and b
    E. None of the above

Laser Power Supplied and Wall-plug Efficiency

75. Slope efficiency _____ threshold.
    A. is a ratio of the output to input energy or power before
    B. reveals the fraction of energy input that emerges as laser output above/past
    C. a or b
    D. None pf the above

76. Slope Efficiency is the same as _____ Efficiency
    A. Conversion
    B. Electrical-to-optical
    C. Wall-plug
    D. All the above
    E. None of the above

77. Threshold is when a laser _____.
    A. starts to output
    B. when steady-state has been reached
    C. a and b
    D. None of the above

78. After turning a laser on and before threshold a laser _____.
    A. does not consume any energy
    B. consumes energy
    C. does not output
    D. b and c
    E. None of the above

79. Electrical signals drawn by power supplies from electrical outlets AC convert the signals to _____.
    A. DC
    B. AC/DC
    C. Any of the above

80. Electrical signals drawn by power supplies from electrical AC outlets can _____.
    A. be of any AC Voltage
    B. only be 120 VAC
    C. only be 240 VAC
    D. a and b
    E. None of the above

81. Laser useful life-time information _____.
    A. is always 5, 000 hours
    B. can be predicted/estimated
    C. is generally, or may be, known by a laser manufacturer
    D. b and c
    E. None of the above

82. Laser lifetime tests are conducted using _____.
    A. Accelerated Aging Tests
    B. Pump Energy Aging Tests
    C. a and b
    D. None of the above

Relative Stability Fluctuations

83. Laser beam performance parameter fluctuations general exhibit a _____ distribution
    a. Normal
    b. Bell
    c. Gaussian
    d. All the above
    e. None of the above

84. _____ level(s) of laser beam performance parameter fluctuation will be acceptable for any application.
    A. Some
    B. No
    C. All

85. The larger the standard deviation the _____ precise a measurement/parameter value
    A. more
    B. less
    C. any of the above
    D. None of the above

86. Relative Stability Fluctuation discloses a specific parameter's performance _____ in the laser output.
    A. precision
    B. average
    C. standard deviation
    D. all the above
    E. None of the above

87. Relative Stability Fluctuations of laser performance parameters _____ be disclosed to the user/buyer by a laser manufacturer unless otherwise requested
    A. will
    B. will not

Threshold, Slope Efficiency, and Warm-up Time

88. Threshold is when a laser _____.
    A. starts to output
    B. when steady-state has been reached
    C. a and b
    D. None of the above

89. After turning a laser on and before threshold a laser _____.
    A. does not consume any energy
    B. consumes energy
    C. does not output
    D. b and c
    E. None of the above

90. Warm-up time includes _____ time(s).
    A. small signal gain
    B. saturation
    C. steady-state
    D. a and b
    E. None of the above

91. Slope efficiency _____ threshold.
    A. is a ratio of the output to input energy or power before
    B. reveals the fraction of energy input that emerges as laser output above
    C. a or b
    D. None pf the above

92. Slope Efficiency is the same as _____ Efficiency
    A. Conversion
    B. Electrical-to-optical
    C. Wall-plug
    D. All the above
    E. None of the above

Electrical Outlets, Power Supplies, and Wall Plug Efficiency

93. Electrical signals drawn by power supplies from electrical AC outlets convert the signals to _____.
    A. DC
    B. AC/DC
    C. Any of the above
    D. None of the above

94. Electrical signals drawn by power supplies from electrical AC outlets can _____.
    A. be of any AC Voltage
    B. only be 120 VAC
    C. only be 240 VAC
    D. a and b
    E. None of the above

95. Power Supply components include _____.
    A. Transformers
    B. Bridge rectifiers
    C. a and b
    D. None of the above

96. Lasers are run using _____ electricity
    A. AC
    B. DC
    C. Any of the above
    D. None of the above

97. Wall-plug Efficiency is the ratio of _____.
    A. Input electrical power to output optical power
    B. Output optical power to input electrical power
    C. Any of the above
    D. None of the above

Useful Lifetime and Footprint

98. Laser useful life-time information _____.
    A. is always 5, 000 hours
    B. can be predicted/estimated
    C. is generally, or may be, known by a laser manufacturer
    D. b and c
    E. None of the above

99. Laser lifetime tests are conducted using _____.
    A. Accelerated Aging Tests
    B. Pump Energy Aging Tests
    C. a and b
    D. None of the above

100. If a laser is to produce powers more than a kilowatt it must _____
    A. weight at least 10 pounds
    B. have a footprint at least 5 square feet
    C. All the above
    D. None of the above

101. Laser lifetime knowledge is useful to their users for _____.
    A. estimating the Return on Investment (ROI)
    B. forecasting and budgeting
    C. a and b
    D. None of the above

102. Laser useful life-time information is made available to the user/buyer _____.
    A. only upon request
    B. upon purchase
    C. through a court ruling
    None of the above

The END!

A graded version of this test is available online at laserpronet.com. Click on Laser IQ Scan to access the test.

Also consider taking this Tech 1 test
FREE Laser IQ Scan/Test

Your score, percentile etc. will be emailed to you and/or your manager instantly/as soon as you compete your test.

laserpronet.com
Empowering the Laser Workforce

Lab Course Final Project Title Page

Company Name and Department

Laser Fabrication and Factory-level Tuning and Performance Test Procedures

by
Type Your Full Name
Month and Year

Written as partial fulfillment of course lab requirements

Table of Contents

Topic   Page

1. Laser resonant cavity fabrication.
2. Laser pump output wavelength tuning
3. Threshold, slope efficiency, and warm-up Time
4. Collimation of diverging Laser Beams
5. Laser Threshold and slope efficiency
6. Laser beam harmonics: temp-tuning and conversion efficiency
7. 8. Laser power supples and wall-plug efficiency

When done
1. E-mail an electronic copy
2. Hand in a bound hardcopy
   of procedures to the instructor on the stipulated due date(s).
3. Get ready for a hands-on exam